Parts of a Whole

Written by Margie Burton, Cathy French, and Tammy Jones

One day, I got five
bananas. I took one
banana from the bunch.
There were now four left.

At my party, we got a big pizza. We cut the pizza into ten slices. I ate two slices and my friends ate three slices.

WIBC ABC
APPROVED

I went bowling. The ball hit three pins. Seven pins were left.

I had a candy bar.
My candy bar had
three pieces. I ate
two pieces and gave
one piece away.

My mom made one big pie. She cut the pie into six slices. Three of the slices had whipped cream on them. Three slices did not.

I had some crayons.
There were eight crayons in the
box. I took out two crayons.
There were six left in the box.

NON-TOXIC
8
• safe and non-toxic, with glitter made of plastic,
not metal.
• specially formulated so the glitter sticks to the
paper. During normal use, glitter particles may
build up leaving clumps that could rub off onto
skin or household surfaces. Soap and water
removes glitter from most surfaces.
For Best Results:
Crayola Glitter Crayons in a cool place
sunlight.
any glitter product,
is advisable.

We got twelve eggs at
the store. I took three eggs
out to cook. We
had nine eggs left
to cook later.

WELCOME TO EGG•LAND'S BEST
If you love eggs, you're in for a treat. We feed our hens a special diet to bring you the best tasting, nutritious eggs we possibly can.
Our hens eat no animal fat. Their all vegetarian diet includes natural grains, canola oil, and a unique feed supplement with Vitamin E.
The result is quite possibly the best tasting eggs you've ever eaten. In fact, one Eggland's Best egg provides 25% (7.5 IU) of the Daily Value for Vitamin E compared to only 4% (1.1 IU) for an ordinary egg.
Only eggs meeting our exacting standards are stamped EB. This is your assurance of the finest EGG•LAND'S BEST quality, flavor and nutrition. (LARGE)
Calories 70
Total Fat 4.5g
Protein 6g
If you are concerned about cholesterol, follow a low fat diet and ask your Doctor or Dietitian about Eggland's Best's Clinical Studies.

One scoop of my ice cream cone fell off! What did I have left to eat?